Designed and produced by
Aladdin Books Ltd
70 Old Compton Street
London W1

First published in the United States in 1987 by
Franklin Watts
387 Park Avenue South
New York NY 10016

ISBN 0 531 10433 8

Library of Congress Catalog
Card Number: 87 50597

Printed in Belgium

The publishers wish to thank those involved in producing this book, especially Dr John Fry, medical doctor and author

Design David West
Children's Book Design

Editor Steve Parker

Researcher Cecilia Weston-Baker

Illustrator Peter Harper

Contents

UNDERSTANDING DRUGS

ALCOHOL

Rob Stepney

FRANKLIN WATTS
New York · London · Toronto · Sydney

hacker Bier himmel der Bayern

INTRODUCTION

Let's have a drink. At a party, at a disco, after work. On a picnic, at lunch, during dinner, at the weekend. To celebrate a birthday, or passing exams, or finishing school. To relax after a hard day . . . any time seems like a good time for having a drink.

"Drinking" usually means "drinking alcohol." For many people, having a drink can become a regular event. Perhaps every day. Bars, many restaurants, clubs, liquor stores – they all sell alcohol. People talk about drinking and about how much they can tolerate. They compare fine wines, good beer and expensive brandies. They drink to relax, to get confidence, for fun, to loosen up . . .

But this familiarity does not mean alcohol is safe. Far from it. Alcohol is a powerful drug with many effects on the body and mind. The effects can be dangerous and harmful, especially if a little too much is drunk over a period of many years. A few hundred people die each year from solvent abuse, and many more from overdosing on narcotic drugs. But tens of thousands die while they are young because of alcohol.

Alcohol is our society's "chosen" drug. It has a special position, which unfortunately makes us less able to see the harm it can do. With such a widespread and accepted drug,

Drinking is a national pastime in many countries.

understanding is all the more important. This book gives the facts about alcohol – its effects on the mind and body, how it is made and measured, why people drink it, and how we can make sure that we use alcohol carefully and wisely, if we use it at all.

When taken in small quantities, even over many years, alcohol can pose almost no risk to health. It is the way we use this drug that's important. Drinking it can be seen as a kind of skill. The way we learn to drink as we grow up affects our ability to handle alcohol for the rest of our lives.

There is another reason for understanding alcohol. This drug has a big influence on what people do and on the way society works. Understanding alcohol is often important in understanding people.

ALCOHOL SLANG

Booze . . . demon drink . . . a quick one . . . what's yours? . . . it must be his shout . . . cheers . . . bottoms up . . . good health . . . one for the road . . .

A few too many . . . getting a bit merry . . . getting happy . . . on a bender . . . tanked up . . . legless . . . smashed . . . out of your skull . . . plastered . . . what a session!

Drunken behavior . . . drunk and disorderly . . . breach of the peace . . . fined, and report to the local police . . . driving home? sure, I'm too drunk to walk! ha ha . . . under the influence . . . fined and banned from driving . . . sent to prison . . .

WHY DO PEOPLE DRINK?

"Looking back, it began when I used to get home from work and pour a stiff drink"

Alcohol is an accepted part of everyday life. Having a drink is part of the celebrations around births, marriages and deaths. Alcohol marks our entry into the world and sees us out of it. It's there at every important event in between. Why do so many people take this drug?

There is no doubt that alcohol brings pleasure to millions of people. It is part of being with other people, a way of relaxing. Taking alcohol seems the "natural" thing to do while talking, or eating or relaxing with friends or making guests welcome. Until recently society seemed to think that there was something slightly odd about people who never drink.

> ***"Jim always was a bit strange. Like, he never touched alcohol. Imagine . . ."***

Going into more detail, there are probably as many reasons for having a drink as there are drinkers.

One of the crowd

Some people drink alcohol simply because other people are drinking it too. At a party or in a bar it is not acceptable to refuse a drink. The problem is, people react in different ways to a certain amount of alcohol. It depends on the time of day, how much you've had to eat, your mood, and how accustomed you are to drinking. After three glasses of wine, say, one person might be hardly affected while another is well on the way to being drunk.

This is why, if you do drink, it is so important to know

The two sides of alcohol . . . happy and merry, and sad and depressed.

your own limits, and to recognize the effect alcohol is having on you, so that you can stop drinking in good time. "Saying no" doesn't only apply to "hard" drugs like heroin.

Courage and confidence

We drink because we think alcohol gives us courage. The way it affects the brain makes many people act more confidently. They feel more like talking to others, more like chatting and making friends.

Occasionally this confidence could seem useful. But alcohol releases inhibitions in all sorts of ways. Under its influence, we might say things we don't really mean, which we regret later. Anyway, isn't it better if the self-confidence comes from inside, as part of the real person, rather than from a chemical that alters the workings of the brain?

"He just went up to the teacher and swore, just like that. What a laugh. Still, he got banned for the rest of the term."

Rest and relaxation

How many times do we see, on TV and in real life, someone reaching for a bottle and glass in order to relax? Perhaps it is after a hard day at work or a big event like a job interview.

Yet there are plenty of other ways to relax and calm down. Listening to a favorite piece of music, or reading, or drawing all may relax us. Exercise, like going for a run or working out, is a good way to "empty the mind" and forget about worries and hassles. Or people can simply sit and

Alcohol can "unlock the primitive" and lead to trouble.

think or perhaps learn breathing techniques and meditation from a book or class.

"Looking back, it began when I used to get home from work and pour a stiff drink. Every night, til it got to be automatic. Then two drinks . . ."

The costs

Making and drinking alcohol (except for making spirits) is perfectly legal in most societies. In the United States, on average, every man and woman drinks about three gallons of pure alcohol each year. In some countries, people spend about a tenth of the wealth created by the country each year on buying alcohol.

Picking up the habit . . . children often copy their parents.

For some people drinking is a major outlay in their weekly budget. For a few, drinks come before healthy food or new clothes when money is short.

There are rules about how old you can be before you go into bars. But for most people for most of the time, the law plays no part.

Yet the costs are enormous. All you need to do is add up the money spent on hospital beds occupied by people with alcohol-related diseases, and all the psychiatrists treating alcoholism, and all the property damaged by people who are drunk, and so on. The bill comes to millions of dollars. And that is before we start thinking of the grief and suffering caused to people whose families are split apart or whose loved ones are killed or maimed by drunken drivers.

Politics, business and drinking

Alcohol use is on the increase in Western countries. People ask the reasons why. Is it the breakdown in family life? Is it because fewer people go to church? Does drink simply go "hand in hand" with modern civilization?

One thing is clear. Money plays a part. The real price of alcohol (taking into account inflation) is going down, and the amount of alcohol drunk is going up. This simple equation gives governments a hold on the alcohol problem, if they want to use it. Taxes can be increased as a way of keeping the amount drunk under control.

But governments haven't been keen to use the tax weapon. This is not because they would lose money – less liquor would be sold, but they would make more in tax on each bottle. A more likely reason is that liquor taxes are

The "glamor" sports help to give alcohol a racy image.

unpopular with voters. Also, the price of alcohol is often included in the cost of living. So governments who want to keep inflation down don't like to raise alcohol taxes.

In addition, there is probably pressure from the people who make and sell liquor. It is big business, with big profits to be made and jobs at stake.

The image of alcohol

Part of the problem is that our society still gives alcohol too good an image. Successful characters on TV and in the movies always seem to be drinking. When we do see people damaged by alcohol, they tend to be the same old characters, tramps and the homeless. These are people who are comfortably distant from the way we see ourselves. They are people who don't make us think too hard about where our own drinking might be taking us.

WHAT IS ALCOHOL?

". . . if I started drinking spirits, it'd be more likely that I would become an alcoholic."

Alcohol is a chemical. In chemistry, the word "alcohol" is used for a whole group of substances. Many of them are extremely poisonous and are used in industry to dissolve other substances. There's methyl alcohol, for example, or "wood alcohol." When taken into the body, even small amounts cause serious effects such as blindness, damage to the brain, and death.

When we talk about drinking alcohol, however, we usually mean ethyl alcohol or ethanol (CH_3CH_2OH). It is made by microscopic living things called yeasts, which live in water and are fungi (related to mushrooms and toadstools).

Fermentation

Yeasts make alcohol by the process of fermentation. In this simple chemical reaction, the yeasts are put in a watery mixture with sugar. They take the sugar and the oxygen from the air, and turn them into alcohol and carbon dioxide. Carbon dioxide is a gas and it makes the mixture bubble and froth. The alcohol stays dissolved in the water. Thinking of alcohol as yeast excrement doesn't make drinking it seem very attractive, but that's what it is.

> ***"What can you expect from a species that drinks yeast excrement until it makes them senseless? (Alien in science fiction story.)"***

Unless we deliberately stop it, the sugar ferments until the rising amounts of alcohol kill the yeasts. The yeasts become drunk, in fact. This is a reminder that alcohol is poisonous to living things.

The strength of a drink

Alcoholic drinks are not pure alcohol. They are mostly water with alcohol dissolved in it. There are a few other chemicals as well and these give different drinks their characteristic smells and tastes and colors.

ALCOHOLIC DRINKS

Name of drink	How it's made	Alcoholic strength
Beer	*Fermented from malted barley, with hops for flavor*	**4 to 6%**
Cider	*Fermented from apples*	**4 to 6%**
Lager	*As for beer, stored and "mellowed"*	**4 to 6%**
Wine	*Fermented from grapes*	**8 to 13%**
Sherry Port Vermouth	*Fortified wines, made by blending wines and spirits*	**15 to 20%**
Liqueurs	*Spirits and flavorings, juices*	**20 to 25%**
Whisky	*Distilled from fermented grain (usually barley) or corn*	**most spirits are about 37 to 50% alcohol**
Vodka	*Distilled from fermented grain, beet, potato or other starchy plant*	
Brandy	*Distilled from fermented grape juice*	
Rum	*Distilled from fermented sugar cane*	
Gin	*Distilled from fermented grains, flavoured with juniper berries*	

Whatever the drink, from a weak beer to the strongest whisky, the kind of alcohol in it is always ethyl alcohol (ethanol). What makes the difference is how much alcohol there is compared to the water – in other words, the concentration of alcohol in a drink.

"Strong" drinks like whisky and gin have more alcohol in them, in proportion, compared to "weak" drinks like ordinary beer. How drunk you get, and the long-term damage you do to your health, depends on how much alcohol you drink. It doesn't matter if you drink it in the form of beer, wine or spirits. It's the total amount that counts.

> ***"I always stuck to beer. I thought that if I started drinking hard liquor, it'd be more likely that I would become an alcoholic."***

Making alcoholic drinks

Yeasts are present on the skins of grapes, as a fine powder called a "bloom." When the grapes are crushed, the yeasts mix with the sugary juice from inside the grapes and fermentation starts. It continues until about 12 per cent of the mixture is alcohol (that is, about one part of alcohol to eight parts of water). Then the yeast dies from alcohol poisoning. This marks the natural limit of the strength of wine.

Beers are made from a mixture of yeast, malted barley and hops. The hops give flavor and the "bitter" taste, and

Breweries and distilleries attract many visitors on trips and tours.

GRUNDLAGDE CARLSBERG

they also help the beer to keep longer. "Malted" barley is barley grains that have been soaked in water and then quickly dried. The sticky sugars from inside the barley are fermented by the yeast. The carbon dioxide made by the yeast forms as gas bubbles and gives some beers their fizzy nature.

Lagers are made in a similar way to beers. They are lighter in colour and are usually stored at a low temperature for several weeks or months to "mellow".

Spirits such as gin, whisky and vodka are made from wine or beer-like liquids by distillation. This involves heating the liquid carefully. The alcohol boils off first as a vapor

Ripe grapes wait to be transported from vineyard to press.

which is collected and cooled and condenses back into alcohol liquid. This liquid is much stronger, and in fact can be almost 100 per cent alcohol.

Distillation sounds simple, but it is in fact very tricky. Besides the usual ethyl alcohol, other alcohols may be formed by heating. If these find their way into the drink, people may become blind and even die. This is why home-made distilling equipment (a "still") for making "moonshine" or other illicit liquor is so dangerous.

> ***"Josey drank the 'shine, at first she felt just fine, but then she said 'My head!,' and then she fell down dead.***
> ***(Folk song.)"***

The tastes of drinks

Many people find alcoholic drinks unpleasant at first. With experience we come to drink the ones we like. In strong drinks you can probably taste the alcohol itself. However most of the taste comes from other chemicals, such as from the particular grapes used in a wine, or the special water used for a whisky.

Beer, brandy and whisky have a lot of these extra, strong-tasting substances. Vodka has very few, and it is quite pure as drinks go, being almost entirely alcohol and water.

Some people become very concerned about the taste and smell ("bouquet") of various drinks, especially wines, whiskys, brandies and beers. The manufacturers build up amazing reputations, and there is a whole industry devoted to discussing different types of drinks, writing and reading about them, describing them and collecting them.

So taste is important when drinking. But not if we have more than one or two drinks. After this, our taste buds lose their sensitivity and become "numbed." And in any case, after a drink or two, alcohol begins to affect the brain, and taste goes out the window.

THE EFFECTS OF ALCOHOL

"It was a real scream. We knocked over a few bushes and broke a car window . . ."

Alcohol affects the body's nervous system and especially the brain. If it didn't have this effect, then we probably wouldn't take it. Alcohol is classed as a depressant drug. So how can it produce the lively, sociable, extroverted behavior we see in someone who's had a couple of drinks?

Alcohol as a depressant drug

Alcohol "subdues" the brain, reducing worries and anxiety. No one knows quite why it makes us feel happy and merry.

> ***"Just one drink and I began to come out of myself. I felt equal to everyone in the room."***

One possibility is that alcohol quiets down the "higher" parts of the brain, in the top part of the skull, more than it affects the "lower" parts of the brain, which join to the top of the spinal cord. The higher parts deal with intelligent thought, memory, speech and the distinctly "human" parts of our behavior. The lower parts deal with more basic aspects such as hunger, aggression and sex.

It is thought that normally the higher parts of the brain keep the lower parts in check. Alcohol depresses the higher parts first, so that the intentions and "instincts" controlled by the lower parts come through more. The result is less thoughtful, more spontaneous behavior.

As a drug, alcohol has two other important effects. It gives a sense of pleasure, and it kills pain. Before modern anesthetic drugs, some people were given large amounts of alcohol to reduce the pain before having an operation or a tooth removed.

Stages in getting drunk

With small amounts of alcohol, the effects are mild. Worries seem to fade and we feel more confident. The people we are with are like good friends, and jokes and laughter come easily. This is alcohol acting as a "social lubricant," helping people to get along and have fun together.

As the drinks continue to flow, inhibitions disappear and we do things we're normally too embarrassed to do. We follow impulses, but usually no harm is done.

> ***"It was a real scream. We knocked over a few bushes and broke a car window . . . nothing serious."***

With a little more alcohol we lose coordination, our balance is unsure and we are clumsy, knocking over drinks or dropping things. Our judgment is less sure. At times, the situation gets out of hand. We act stupidly, not thinking about the consequences or ignoring them. Our normal sense of right and wrong is distorted, and people in a group may become aggressive. Fights start, property is smashed and there may be trouble with the police.

Speech and memory

The "higher" parts of the brain are concerned with speech, memory and understanding. Alcohol affects all these abilities. We stumble over words and try to figure out what

One: a social lubricant.

Two: taking the brake off behavior.

someone else has said. We can't remember names or places. This starts to happen at alcohol levels which are about halfway to causing unconsciousness, so it is a serious warning sign.

If we continue to drink, and we lose awareness itself and fall into a drunken stupor, or "coma." This is probably fortunate, for alcohol levels are by now well on the way to paralyzing the part of the brain needed to keep us breathing.

The place and mood

The effects of alcohol also depend partly on us. As with other drugs, what alcohol does depends partly on the situation and the mood at the time.

Usually, if we are happy, alcohol will make us feel even happier. If we're sad and depressed, alcohol will sink us

Three: trouble seeing and talking.

Four: "crashed out," sleeping it off.

deeper into gloom. The idea of people "drinking their worries away" or "drinking to forget" does not really work. Often they feel even worse.

> ***"After Keith left, she went straight out to a bar. When she got back she was smashed. She just burst into tears and cried for hours."***

Drinking in a noisy and lively atmosphere is likely to make someone even louder and more rowdy. Having a quiet drink with friends over a meal is unlikely to have this effect.

Society itself has different ideas about the effects of alcohol. In some parts of the world people expect alcohol to produce laughter, while others expect sadness and tears.

For some cultures being drunk is amusing (provided it is not too drunk), while for others it is a matter of deep shame.

Getting alcohol into the body

It is alcohol's concentration in the blood that matters, since the more that gets into the blood, the more that will pass into the brain and other vital organs. As we drink the stomach and intestines absorb alcohol and it starts to appear in the blood within three to five minutes.

Blood alcohol levels usually rise faster when we drink liquor than when we drink the same amount of wine or beer. They also tend to reach a higher, sharper peak. This means the drug's effects on the way we feel may be slightly different.

Eating and drinking

Having food in the digestive system slows absorption of alcohol. This will delay its effects, but it won't prevent them. With an average meal in the stomach and intestines, alcohol levels from a single drink are highest in the blood about 30 to 45 minutes after it is taken.

Getting rid of alcohol

Even if we never drank alcohol, we would still have a little in our bodies. This is because microscopic organisms in our intestine naturally ferment a little of the sugar we eat. Perhaps because of this, our bodies deal with alcohol quite efficiently.

Most of the alcohol we drink is broken down into other,

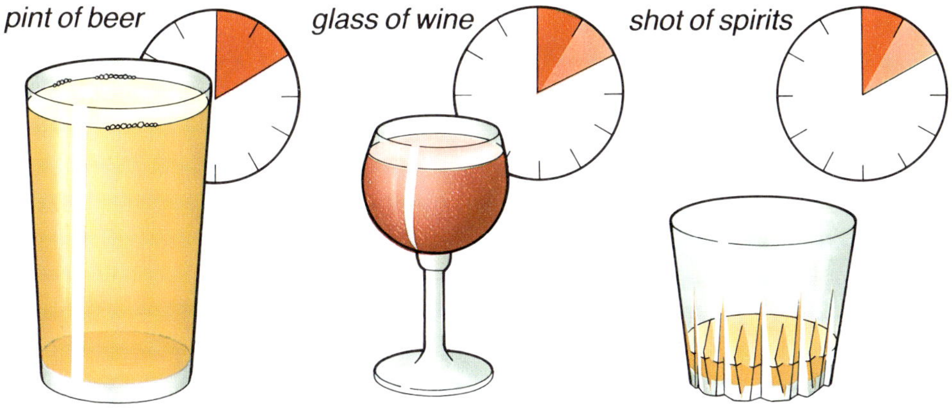

The alcohol in drinks takes some time to be removed from the body.

less harmful substances. This work is done by body chemicals called enzymes, most of which come from the liver. On average, it takes about an hour for the body to break down one "drink" (eight grams) of alcohol. So that's about two hours for a bottle of beer and one to two hours for a glass of wine or a mixed drink.

Tolerance – becoming used to alcohol

The first time a person drinks alcohol, one or two drinks will probably make him or her drunk. But anyone who drinks regularly soon finds he or she can take more and more without getting drunk. Alternatively the same amount of alcohol affects the person less and less. "Experienced" drinkers take amounts of alcohol that would make a newcomer fall dead drunk (if not dead) under the table. This "getting used to" a drug is called tolerance.

The more we drink, the more enzyme the body makes to break down the alcohol into less harmful substances. This is one of the most important reasons the body comes to tolerate alcohol.

The opposite also happens. People who have not been drinking alcohol for a while, perhaps because they've been ill, lose their tolerance. They have to take things very slowly if they start drinking again.

Another reason why tolerance develops is that we learn what to expect and how to control the effects of alcohol. We know what it is going to do to us, and we can manage the consequences better.

Tolerance also differs from one person to another. Some of us are probably born with a more efficient alcohol-disposal system than others.

Storing up problems?

The substances which alcohol is broken down into may not be completely harmless. Some of the chemicals produced may lead to a kind of arthritis, called gout. Others increase the amounts of fat in the blood and liver. Having a "fatty" liver is part way to cirrhosis of the liver – probably the most dangerous disease caused by alcohol.

Alcohol and other drugs

People who drink a lot are also likely to smoke cigarettes. Use of the two drugs goes closely together. So do their effects on health. Cigarettes on their own can cause cancer of the mouth and throat, but this danger is even greater in someone who drinks heavily as well.

An extra risk with alcohol and other drugs is the danger of an overdose. Alcohol is a strong poison, but it doesn't often kill directly. Before people get that far they usually feel sick and vomit, in this way getting rid of some of the alcohol. Rarely do people choke on vomit when they are unconscious after a drinking session.

Much more common is death due to mixing alcohol with other drugs. Taking alcohol with barbiturates is very dangerous. This goes for any sleeping pills, sedatives or tranquilizers. Mixing alcohol with hard drugs such as heroin is a recipe for disaster.

OLIVIA

- **She took heroin through tube**
- **She ordered a pint of sherry**

Alcohol and other drugs can be a fatal combination.

"Keith had always drunk a lot, but when he began doing heroin as well, that's when we got really worried . . . "

Does black coffee sober you up?

People say black coffee makes you sober. It certainly does not do that. Nor does it lower the amount of alcohol in the blood, or help the body to get rid of alcohol faster.

However, to a small extent, coffee may counteract how you feel after drinking. Because caffeine (the main drug in coffee) is a weak stimulant of the brain, and alcohol is a depressant drug, one may partly balance the other. But any effect is very weak.

In fact there are hardly any substances that speed the removal of alcohol from the body. Fruit sugar (fructose) may, but you'd probably have to eat so much of it that you'd be sick. Once alcohol is in the system, you have to wait patiently until your body gets rid of it. And waiting patiently means not taking any chances – it takes one to two hours for the body to get rid of one glass of wine.

Drinking and driving

Driving safely – and doing any other task that involves judgment and coordination – is affected by alcohol. Even after one or two drinks, a driver does not react as fast. His risks of accident and injury go up fast. The same goes for people operating machinery or power tools or those involved in dangerous sports such as skiing or shooting.

The really dangerous thing is that people do worse at these tasks – even though they feel they're doing better. It is an entirely false feeling. Scientists know that even people below the legal limit of alcohol for driving are not in full normal control. Even drivers who have had a drink and are under the legal limit may still be affected by the alcohol and

Yet another Saturday night smash, the result of drinking.

have an increased chance of accidents.

Between one third and one half of all drivers killed in road accidents have high levels of blood alcohol. Thousands of bystanders, pedestrians and other motorists are killed or injured because of drunken drivers. Most societies are beginning to understand that no one who drinks should drive as well. If they do and they kill someone, it's the same as murder.

DAMAGE THAT CAN BE DONE

“When I think back, a lot of his temper was caused by drinking.”

7
8
9
10
11
12

A scruffy, homeless old man sits on a park bench surrounded by empty bottles. He is begging for money to buy a drink. When we think of the harm done by alcohol, this is probably the image we have.

Men and women who are like that are often alcoholics. But there are many perfectly respectable people who depend just as much on alcohol to get them through the day.

Who are the alcoholics?

Business people with a bottle or two of whisky tucked away in their desks, journalists who are never out of the bar, factory workers who keep their liquor in the locker room and spend all night in the pub, parents stuck at home who can't get by without frequent trips to the drinks cabinet . . . these are all problem drinkers.

They all depend on alcohol. Their bodies can't function without it. They may not fit the popular picture of an alcoholic, but the important thing to realize is that alcohol can do damage even to people who aren't alcoholics. Heavy drinking can cause work and family difficulties, as well as health problems.

> ***"He was always in the bar, always slightly drunk. You could have a good chat with him. But he never talked about himself, only about the TV or other people in the place."***

Under stress, the pressure's on . . . but is liquor the solution?

A sad face of drinking . . . alcohol abuse can lead to child abuse.

Social and family problems

When heavy drinkers are late for work or when they start to do badly at their jobs they may lose their jobs. Perhaps it's because they are still drunk from the night before or they have a hangover. This can happen at any age. And once someone gets a reputation as a problem drinker, other jobs can be hard to find.

Families suffer badly when a member is a heavy drinker. To these people, time in the bar is more important than time at home. Money is short. Wives are beaten by men who come home drunk. Children are treated badly. Marriages break up.

Other problems often play a part in family difficulties. The home may be crowded to begin with. Or the husband or wife loses his or her job. We can understand why they turn to alcohol, but it provides no solution. It often makes matters worse.

“When I think back, a lot of his temper was caused by drinking. The first baby cried a lot and he used to go out in the evening to a bar, to get away. He came back one night, woke the baby, and got into a real rage . . . I was scared.”

Arguments and fights caused through drinking may bring problems with the police. Half of all murders take place when the attacker has been drinking. Some people turn to theft to get enough money to pay for all the alcohol they need. An accountant cheats clients and a tramp shoplifts a bottle of vodka. Alcohol and crime are linked at all levels of society.

So stupid. I hung around at the back of the store. When he went into the back room I tried to open the till but it got stuck. I forced it and he came out. How dumb can you be?”

The hangover

Hangovers and headaches from “the morning after” are a gentle, short-term reminder that alcohol is a long-term poison. As the amount of alcohol consumed each day goes up, so do the risks of disease for many parts of the body.

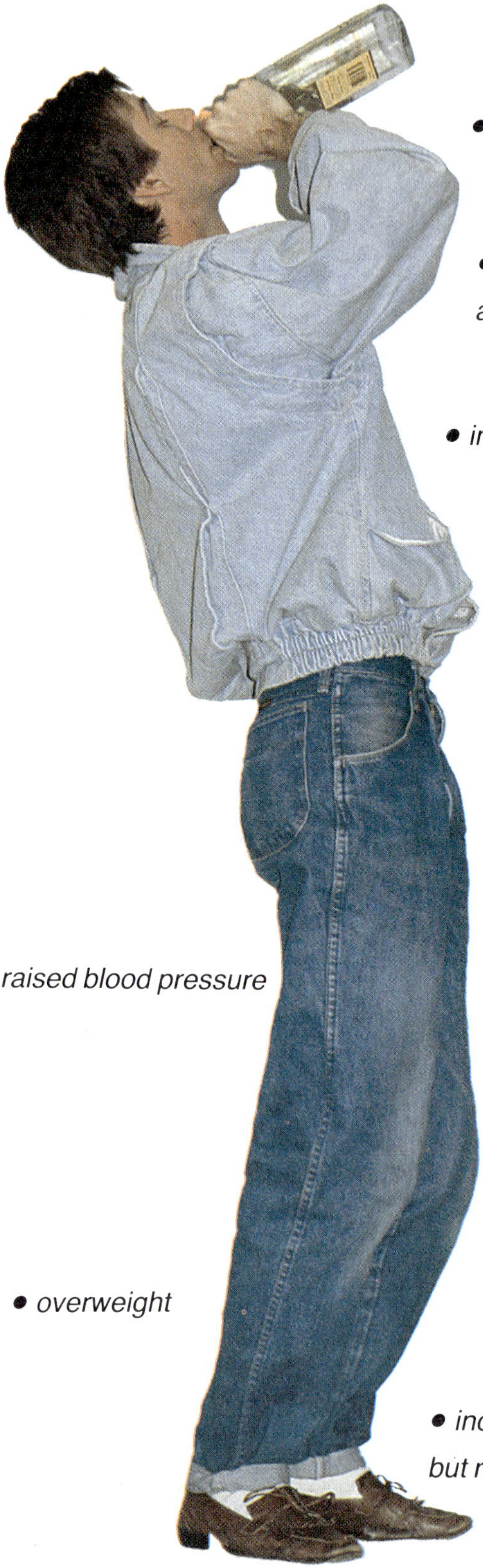

• headaches and hangovers
• loss of memory and mental powers
• increased risk of heart attack
• damaged heart muscle
• stomach irritation, ulcers
• liver cancer
• raised blood pressure
• cirrhosis of the liver
• overweight
• increased sexual desire, but reduced performance

People often ask: "What's a good hangover cure? A raw egg, or a strong cup of coffee?" When you suffer from hangovers enough to ask, there is only one cure: cut down on the drinking.

Liver

Liver problems are the most common result of alcohol abuse. The liver grows in size and develops deposits of fat. In itself this is not too dangerous and the condition may clear up in a few months in people who stop drinking.

But many drinkers move on to more serious harm. Orderly rows of healthy liver cells are replaced by unhealthy, disorganized ones. The liver swells, then gradually hardens and shrinks. Blood can no longer pass through it. Poisons that are usually dealt with by the liver, including alcohol itself, build up in the body.

This is cirrhosis of the liver, a disease that is often fatal. The only way to stop it worsening is to stop drinking completely, forever.

Alcohol may also increase the risk of liver cancer.

Heart

Alcohol seems to speed the heart rate a little and increase blood pressure. It is a common cause of blood pressure problems. Heavy alcohol use may damage the heart. The small arteries supplying the heart muscle with blood are more likely to become blocked, increasing the risk of heart attack. And alcohol may damage the heart muscle directly.

Alcohol may have many wide-ranging effects on the body.

Stomach

Too much alcohol irritates the stomach. The lining becomes inflamed, and ulcers may develop. It is difficult to digest and absorb food properly. This may lead to terrible stomach pains and signs of malnutrition.

Brain

Alcohol is a cause of a condition called dementia – literally "without a mind." Your memory goes. It's impossible to concentrate. In advanced stages, people become not much more than vegetables.

> ***"There was a time which I can't place at all. Susan said I never came home til after midnight, then I got up late and went straight out. I think I just wandered about, buying booze from liquor stores, sitting in the car and drinking it. What a state!"***

Sex

Alcohol can increase the desire to have sex. But it does nothing much for performance. And alcohol also does longer-term damage. People who drink heavily are less fertile, and sexual capacity falls.

Alcohol and the unborn child

Babies born to mothers who drink heavily during pregnancy are more likely to be deformed and to develop more slowly than other babies. This may be partly because mothers who

The calories in alcohol can translate into inches round the waist.

drink a lot do not eat well. They may also be eating irregularly or even missing meals.

Drinking early in pregnancy is more likely to cause problems than drinking later on. Anyone who is pregnant, or wishing to become pregnant, should drink only very little, if at all.

Alcohol makes you fat

One gram of alcohol contains seven calories. Drinking is such a good source of energy that alcoholics get all the energy they need from their drinks. But they don't get all the vitamins, proteins and other nutrients the body needs. In fact much of the damage alcoholics suffer is due to a mixture of the drink itself and the effects of being undernourished.

If you eat an ordinary amount of food, and drink heavily as well, the extra calories from the alcohol will certainly make you put on weight. People who want to slim down must be very careful not to drink much alcohol.

> ***"He always said it was his job, having to take people out, wining and dining. All that food, and the booze, next time I saw him he looked like a fat man in a cartoon."***

DRINKING OUT OF CONTROL

"If I could get a couple of drinks down I'd be much less tense."

> ***First the man takes a drink. Then the drink takes a drink. Then the drink takes the man. (Japanese proverb.)***

You start off in control. You use alcohol, and you don't let it use you. You take the drink.

Then, for some people, situations occur when they drink too much. It may only be a few occasions. Some people know there are times when they should really say "no" to another glassful, but they say "yes" instead. We can call it the drink talking. The drink takes another drink.

And finally there are some people who are simply destroyed by drink. Often their families are destroyed too. The drink has taken over the people.

The "alcoholic"

The word "alcoholic" sounds as if it has an exact medical meaning, but it doesn't. There is no set amount or level of drinking, like so many bottles of beer or whiskys a day, which defines an alcoholic. You can't say two bottles a day is OK, but three means you're an alcoholic.

The word "alcoholic," however, is a useful shorthand to describe some people who depend, for whatever reason, on alcohol to get them through the day.

Ideas about alcoholism

We do not know why some people come to depend on alcohol and others don't. Until we do know, the only safe

The morning after the night before, fuzzy and bleary-eyed.

approach is to assume that everyone can be a victim of problem drinking.

We know there are differences between people from different groups in the way their bodies handle alcohol. Chinese and Japanese become flushed more quickly than Europeans, for example. It could be that there are also important differences among individuals. If our bodies deal with alcohol in different ways, perhaps some of us start life more likely to become alcoholic.

Some evidence supports this idea. We know that the genes we inherit from our parents have some effect. Children of parents with an alcohol problem are more likely to have an alcohol problem themselves. This is true even if the children are brought up away from their natural parents by people who don't have difficulties with alcohol.

"Alcohol seemed to satisfy some special need I had. It made me feel good. Made me forget my troubles. But that wasn't it. Other people drank for those reasons. And they didn't become alcoholics."

Personality

For a hundred years psychologists have looked to the mind for an explanation of alcohol abuse. They said that alcoholics take to the drug as a way of dealing with conflicts with their parents, for example. They said that alcoholics have a weak personality, or that they are naturally anxious people who drink so they can cope with everyday events. Or that they are people whose ambitions in life have come to nothing.

"I suppose I first discovered that drinking would do something for me at an early age. At 17 or 18 I was desperately shy . . . But then I found that if I could get a couple of drinks down I'd be much less tense."

Some of these ideas ring true. But the personality

Children often want to be grown-up, and drinking is what grown-ups do

features we link with too much drinking may have been caused by alcohol, and not the other way around. Scientific studies that follow hundreds of people from early life through to adulthood have found no good evidence that a particular kind of personality leads to drinking problems later on.

Stress

Stress is difficult to define. But you often find that people who are going through periods of anxiety or worry increase the amount they drink. Divorce and separation, loneliness, losing a job, going bankrupt, the death of a family member or friend – any of these can increase the amount of alcohol drunk.

Most people return to normal once the crisis is over. But having to turn to drink is a bad sign. Some people spiral into heavier and heavier use.

Culture and society

Many religions limit alcohol use to ceremony while others, such as Muslims, Mormons, and strict Jews, prohibit it entirely, and its abuse is never acceptable. This tells us that cultural factors are important. Whatever someone's personality, and even if they fare badly in life, belonging to one of these strict groups can give protection against alcoholism.

Differences between women and men

Women tend to weigh less. They tend to have less body water in which alcohol can be diluted. So, on average, if a man and a woman drank the same amount of alcohol, the woman would have a higher blood alcohol level.

Drinking and driving

The advised upper limits do not apply to driving. Any amount of alcohol is dangerous if you intend to drive. It takes an hour, on average, to get rid of just eight grams of alcohol. So it would take around eight hours to get a bottle of wine out of your system.

If you have a couple of bottles of wine in one evening (or the same amount of alcohol from any other drink), you'll probably still be above the legal alcohol limit for drivers the next morning.

Regular and binge drinking

People who drink regularly, every day, suffer more liver damage than people who drink the same total amount of alcohol, but who drink it over just a few days in a "binge" and then stop for a while. So having a few days off (at least) to rest the liver is better than drinking every day. Unfortunately, people who drink heavily now and again may be more likely to have accidents.

The cost in accidents and crime

At the legal limit for driving (one part alcohol to 1,000 parts blood in most states, or an equivalent concentration in breath or urine), a person is still more likely than a non-drinker to have an accident.

Alcohol is also heavily involved in crime and violence. Experts estimate that drink plays a part in half of all cases of wife battering, three in five cases of child beating, and one in five crimes of violence.

And alcohol causes much lost production in industry and business.

WHAT'S A SAFE LEVEL?

"Coming off booze was like coming out of a dark room . . ."

In Western society, almost everyone has tried an alcoholic drink by the age of 15 or 16. Many have their first drinks before this – usually with their families. During this early period each of us has to develop an attitude toward alcohol. Are you going to drink at all? If so, how often and how much? Whether to drink at all is very personal. But on the question of how much, doctors have important advice to give.

How much?

Some medical experts advise that an adult in perfect health with no history of alcoholism can have two drinks a day – and on occasion three drinks in one day – without serious damage to his or her health. The American Heart Association has even stated that two drinks a day may help circulation. However, alcohol can produce so many side effects that any attempt to set guidelines for "safe" drinking is bound to be controversial.

In May 1987 the New England Journal of Medicine published the results of a study by the Harvard University Medical School and the National Cancer Institute that linked alcohol and breast cancer. Research showed that women who have only one drink a day run a 60 to 100% higher risk of getting breast cancer than women who don't drink at all.

Pregnant women have to be especially careful of alcohol because drinking may risk the health of their unborn children. Some babies born to women who drank have suffered from many problems including low birth weight, deformities and damage to the brain or central nervous system, mental retardation or learning disabilities. These symptoms are called Fetal Alcohol Effects and they could have been prevented if the mothers had not had alcohol during their pregnancies.

> ***"I am David. I am an alcoholic. I've always been an alcoholic. I will always be one. I can't touch alcohol now. It will destroy me."***

Withdrawal and the DTs

People usually drink to feel good. The alcoholic drinks to stop feeling bad. The body is so dependent on alcohol that it can't work properly without this drug. Dependent drinkers keep on drinking to prevent the unpleasant effects caused by the drug not being there. These unpleasant effects are called withdrawal.

In alcohol withdrawal, a person may shake uncontrollably, sweat, panic and have hallucinations. It's a kind of temporary madness, with loss of memory and agitation. The trembling and the madness (delerium) give rise to the name "delirium tremens", or simply DTs.

The hallucinations can be of strange animals, such as insects crawling over the skin. Animal hallucinations underlie the famous idea that alcoholics see "pink elephants" when they can't get a drink.

Dramatic DTs are uncommon. Morning shaking is the most usual sign of alcohol withdrawal. If it's bad enough, you can't get a cup from the saucer to your mouth. A few days off alcohol, though, and the shakes go away. They can

Safe levels: one unit is half a pint of beer or a glass of wine.

often be avoided altogether by gradually cutting down on the amount drunk.

Symptoms of withdrawal are one way of looking at alcoholism. Another is simply to say that an alcoholic is someone who drinks, who has bad problems because of drinking, but who goes on drinking anyway.

Warning signs

People start to think seriously about their drinking for all sorts of reasons. Maybe a relationship has broken up because of it, or drinking has led to a brush with the police. Being hung over in the mornings stops being funny. Other things in life may start to be eased out by time spent drinking or recovering from it. Overweight may become a problem.

All these are signs that drinking is becoming too

regular a habit. If at any time you think these statements fit you, watch out:

- Always thinking about the next time you can drink.
- Lying to youself and others about how much you drink.
- Feeling guilty about drinking.
- Memory blackouts about what you did the night before.

Getting help

If you think you have an alcohol problem, or you're not sure, then talk to someone who knows. Go to your family doctor or to one of the organizations listed on page 60. The talk will be confidential. Alcohol is not a way of solving life's problems, it's a way of adding to them.

> ***"Coming off booze was like coming out of a dark room into the spring sunshine. Everything was alive, colors and sounds."***

A good guide

- **Don't** drink every day.
- **Don't** rely on alcohol to help you cope.
- **Don't** drink on your own.
- **Don't** drink alcohol while taking other drugs. This means any other drug, whether it's been prescribed by a doctor, or bought over-the-counter from a chemist, or bought illegally, or been offered by friends.
- **Don't** drink without some food in your stomach. Eating while you drink will delay the effects of alcohol.

FACTFILE

The "alcohol" in drinks

Chemically, there are many different types of alcohol. The one in alcoholic drinks is ethyl alcohol or ethanol. It is a colorless liquid – the colors of drinks come from other ingredients. Beer is about one part alcohol to 20 parts water. Wine is one part alcohol to between seven and 12 parts water. Most spirits are about one part alcohol to one part water (half and half).

Pure ethyl alcohol, for medical use, is a clear liquid that feels cold on the skin. It is used as a preservative in the laboratory (for "pickling" specimens) and is used to kill germs, such as wiping on the skin with cotton before an injection.

Yeast, the alcohol-makers.

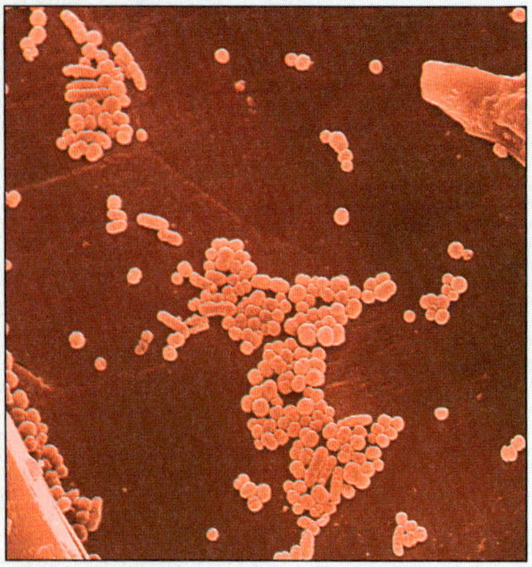

Tell-tale signs of alcohol abuse

If you suspect someone of drinking secretly, or drinking too much, there are several tell-tale signs that include:

- the smell of liquor on the breath, and sometimes the smell of cigarettes as well (many places where liquor is consumed are also frequented by smokers)
- hidden bottles or other more anonymous containers
- signs of drunkenness such as not being able to concentrate, flushed face, slurred speech, unsteady and clumsy movements, loss of memory, drowsiness, violent behavior.
- first to the bar, first to finish a drink, first to buy another, last out

MEASURES OF ALCOHOLIC STRENGTH

%Alcohol	U.S. Proof	U.K. Proof
10	20	11.50
20	40	35
30	60	52.50
40	80	70
41	82	71.75
42	84	73.50
43	86	75.25
50	100	87.50
57.15	114.28	100
60	120	105
70	140	122.50
80	160	140
90	180	157.50
100	200	175

Emergency!

If someone gets into real trouble with alcohol, the person may become unconscious. The best advice is to send or telephone for an ambulance or get some other form of emergency medical care. While waiting, place the person in the recovery position shown below. This way, if he is sick, he is less likely to choke on the vomit.

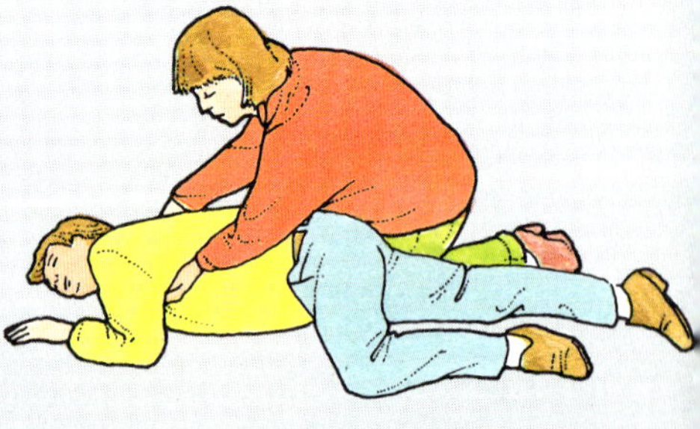

DRUG PROFILE

Common names	Alcohol, booze
Chemical names	Ethyl alcohol, ethanol
Drug type	Depressant
Made from	Fermentation by yeasts of sugars in plants, such as in grapes, grains
Main effects	Feelings of well-being, reduced anxiety, sociability; blocks pain; reduced ability to do tasks that involve coordination and judgment, such as driving; distorted feelings and perceptions; feeling sick and vomiting; headache and lethargy several hours later
Addictiveness	Produces very strong addiction ("alcoholism") in some people. Risks physical and psychological dependence; damage to liver (cirrhosis); damage to stomach; loss of memory
Producer countries	Worldwide except in strict Islamic states such as Saudi Arabia Arabia
Consumer countries	As for producers
Users and abusers	In 1985, there were an estimated 17.9 million adult alcoholics and alcohol abusers in the United States; 4.6 million youths between the ages of 14 and 17 had alcohol problems. In a 1985 study of high school seniors, 5% said they used alcohol daily; 37% said they had had at least 5 drinks at least once in the previous two weeks; and 65.9% said they had used alcohol in the previous 30 days.
Example of cost	Costs vary from state to state, according to local taxes, but on the average a bottle of beer costs $1.00, $3.00 and up (depending on quality) for a bottle of wine, and $5.00 or more for a bottle of liquor.
Legal status	In nearly all the states the drinking age is 21. This means that no one under the age of 21 can purchase a drink in a public place or buy liquor in a store.
Drunkenness	If you are noisy and disorderly, the chances are good that the police may stop you and determine if you are intoxicated. More significant is the very strong campaign against driving while under the influence (DWI). It is illegal to drive with more than a 0.10 blood alcohol level (one part alcohol to 1,000 parts blood) in most states.
Home brewing	It's legal to make up to 200 gallons of wine or beer in your own home each year without a license as long as it's for personal use and not for sale

SOURCES OF HELP

National Hotlines

National Institute of Drug Abuse Treatment Referral
1-(800) 622-H-E-L-P

This hotline is staffed from 9.00am to 3.00am on weekdays and from 12 noon to 3.00am on weekends. Counselors can talk with you, refer you to a drug treatment program, or answer questions about drugs, rehabilitation, health or legal problems.

New York State Division of Substance Abuse Services
1-(800) 522-5353

This toll-free number reaches counselors who can provide referrals for treatment or legal advice, or over-the-telephone crisis intervention.

National Federation of Parents for Drug Free Youth
1-(800) 554-K-I-D-S

This is not a crisis hotline, but a place to call for drug information. This educational organization provides both parents and kids with informational pamphlets, books and videos.

Help and support in stopping drinking

Alcoholics Anonymous
PO Box 459
Grand Central Station
New York, NY 10017
(212) 696 1100

AA is a self-help group founded by and for alcoholics. You do not have to quit drinking for any length of time before you can participate; the only requirement is the desire to stop drinking. Local chapters are listed in the phone book.

National Self Help Clearinghouse
33 W 42nd Street
New York, NY 10036
(212) 840 1259

Can provide information on self-help rehabilitation organizations in your area, or put you in touch with one of the twenty-seven state and local self-help clearinghouses around the country.

For family help

Al-Anon Family Groups
PO Box 862
Midtown Station
New York, NY 10018

A support group for friends and relatives of people with drinking problems. Al-Ateen is a service of Al-Anon specially for younger family members.

Referrals, Information and More

Women's Alcoholism Center
2261 Bryant Street
San Francisco, CA 94110
(415) 282 8900

This center has both a residential and an outpatient facility for alcoholic women and their children. Provides individual and group therapy (including psychotherapy and play therapy) for children, and offers alcohol and drug education.

National Association on Drug Abuse Problems
355 Lexington Avenue
New York, NY 10017
(212) 986 1170

Conducts a drug prevention program and offers family counseling.

Helping Youth Decide
National Association of State Boards of Education
PO Box 1176
Alexandria, Va 22313
(703) 684 4000

Write for their free booklet about making informed decisions concerning alcohol, smoking, drugs and other issues. This organization also organizes parent-student workshops and community projects.

WHAT THE WORDS MEAN

addict someone who needs to keep taking a drug in order to remain "normal" and stave off withdrawal effects on the body and/or mind. "Addiction" has a slightly different meaning to "dependence," in that it usually refers to someone who has been dependent on a drug for some time, and it is more tied up with the person's lifestyle and society's view of him or her. This is especially true of the "alcoholic," usually (and incorrectly) thought of as a tramp or down-and-out.

alcohol in the context of drinking, the chemical ethyl alcohol or ethanol which acts as a depressant

dependence the need to keep taking a drug regularly, either for its effects on the body (to keep away withdrawal symptoms, for instance) or its effects on the mind (such as to make the user think he or she is "getting through the day")

drug any chemical or other substance that changes the body's workings (including the way the person's mind works, his or her behavior, etc)

drug abuse non-medical drug use with harmful effects, on the abuser and possibly on others

drug misuse using drugs in a way which people in general would see as not sensible, or not acceptable, and possibly harmful

DTs (delirium tremens) the strange behavior and trembling due to withdrawal after being dependent on alcohol

intoxication loss of control of body and mind, usually due to a drug; often applied to drunkenness due to alcohol

per cent the amount of pure alcohol by volume, in a drink (the rest is water). For example, 50 per cent means 50 parts of alcohol in 100 parts of drink, that is, half alcohol and half water.

proof An old test in which you poured the drink on gunpowder. If you could still light the gunpowder, it was "proof" that the drink contained a certain proportion of alcohol (alcohol is inflammable). 100 degrees proof equals 50 per cent alcohol in the United States.

specific gravity used mainly for beer, this gravity measures the density (weight for a certain volume) of the drink, compared to alcohol.

tolerance when the body becomes used to a drug, so that the same dose begins to have less effect and increasing doses must be taken for the same effect

"units" used generally to measure alcohol intake. One unit is 10 grams and is roughly equal to half a pint of beer, one glass of wine and one ounce of spirits.

withdrawal the effects on the body and mind when a person suddenly stops taking a drug after being dependent on it. The effects are usually unpleasant and with alcohol may include heart flutters, headaches, sweating, nervousness and edgy behavior, trembling and hallucinations

INDEX

Photographic Credits:

Cover and pages 10-11, 23, 27 (both), 40 and 47: Vanessa Bailey; pages 4,9 (left) and 20: Rex Features; page 7: Art Directors; pages 9 (right), 12,26 (left), 32-33 and 49: Network; page 13: Robert Harding; page 15: Spectrum; page 19: Photosource; pages 26 (right), 35 and 45: Hutchison Library; page 36: Zefa; page 38: John Frost; page 53: Frank Spooner; page 58: Science Photo Library.